Fox

ENCOUNTERS IN THE WILD

JIM CRUMLEY

ONE

HEATHROW AIRPORT in my mind is a loathsome phenomenon without a redeeming feature.

Oh, except this one: I had a London publisher at the time and I would be sent airline tickets to fly from Edinburgh to London for meetings with my editor. I cannot believe that was what the Wright Brothers had in mind when they showed the world how to fly, but it happened a few times over several years in the course of producing and publicising three books. On what would prove to be the final occasion, I was staring mindlessly out of an aircraft window, a helpless unit of its hapless cargo, as we dawdled through the ritual of queuing for the right to take off, and all I could see was the curving procession of aeroplanes in front of us, the grey sprawl of the apron and an apparently unending prairie of short grass, the Slough of Despond wrought in grey concrete and green turf. It is the only place I have ever been where there was no landscape of any kind. Instead, there was just a dead flat dull green stain, a sprawling grass desert that blurred at last into the lowered undercarriage of a bloated grey, bottom-heavy sky that leached a grubby sub-species of drizzle such as a disgruntled God with a vengeful sense of irony might have devised.

All the while, the engines throbbed and wailed, the pitch rising and falling as we inched forward then stopped again and again and yet again, a soundtrack of such intense disagreeableness that it was surely composed specifically to complement that moment of bleak purgatory. I suspected the disgruntled God again, God the vengeful composer.

Then the fox.

It appeared from beneath the aircraft in which I sat, and, this being August, it was as unexpected as Santa Claus. Its indifference to its surroundings was shocking yet somehow miraculous, God the miracle worker. My first thought was uncharitable: either it was one of a race that had evolved a mysterious defence against the maddening, deafening assault of engine decibels, or it was already quite mad and quite deaf.

Away it walked, away and away and away.

Where are you going? What are you doing?

It walked far, far away across the grass the way foxes do, in a dead-straight line, hind foot stepping carefully in the print of the corresponding fore foot, the way wolves do, because in snow it halves the effort of trail-breaking, and somehow, in the absence of snow, they never bothered to dispense with the habit. Somewhere beyond the limit of my vision through an aircraft window veined with blown rain, there must have been a destination, an end of the line for the fox that would reveal its purpose (or its purposelessness), but all that I could see suggested nothing so much as the most pointless journey on earth.

And then I thought: how could it all have come to this for the red fox tribe? Where and why did it all go wrong? And when did the fox strike a bargain with the one species on earth – yours and mine, the people – that has invested so much creative energy into finding ways of killing it… when did it strike a bargain that allowed it to live more or less unmolested in our towns and cities (albeit with a better than average chance of being run over before it is two years old), but not in our fields and forests and moors and mountains where it truly belongs and where its routine fate is to be hunted, shot, trapped, poisoned or ripped apart by mob-handed hounds, after which its tail might be hung from a gamekeeper's gibbet as evidence of his good stewardship of his boss's acres?

There may be ten thousand foxes in London alone. No-one knows for sure (how do you count them?) but students of urban foxes have tried and come up with that best guess, and from the lofty vantage point of my own home base just south of the first of the Highland mountains, I am prepared to believe it. But these are not my idea of foxes, and that bargain between my tribe and the fox tribe is not my idea of a fair deal.

And yes, I know that Heathrow Airport is not, strictly speaking, London (and whether this fox beyond the aircraft window is included as one of the ten thousand I confess I neither know nor care), but again, the view from my northern outpost reinvents London as a vast seepage of greyness and noise, and Heathrow and its fox (its foxes?) are embraced in its unearthly maw. I have watched many foxes disappear, and occasionally the disappearance has been inexplicable (for the fox also has that other wolf trait of how not to be seen when it so chooses), but never, until that moment, had I encountered a fox whose very presence was inexplicable. So I watched it walk on and on, and away and away and without lifting its head or breaking its stride, until at last it dwindled into the last yard of available distance where it was consumed by that grey-green twilight world from which (it seemed to me at that moment) the only possible means of escape was up.

We taxied at last, we turned, we roared, we thundered, we flew. We burst open the sky, we soared into a sunlit Valhalla of vivid blue where we laughed in the face of the Vengeful One, and by then Heathrow was just a bad taste in the mouth, and a whisky from a smiling stewardess took care of that. We flew north to Scotland and sanity, circled over Edinburgh in sunshine and there in the northwest (and tilted weirdly at an angle of about thirty degrees) was the mountain arc of the Highland Edge, with that so-familiar dark blue-green pyramid at its centre, Ben Ledi in its summer clothes, and for so long now my friendly neighbourhood mountain, my fox mountain. We levelled out then and swooped to meet the earth, and the landscape of my life was back on an even keel, except that at the very moment when the aircraft wheels screeched and bit into Scottish soil, the Heathrow fox returned to begin to haunt me. Twenty years later and from time to time, it haunts me still.

Where are you going? What are you doing? Did you never hunt a mountain boulderfield or roll in a fragrant wood of wild hyacinths? Did you ever even know that these things exist?

TWO

JANUARY ON BEN LEDI, centrepiece of the Highland frontier if you approach it from Stirling and the south, first and last mountain of Scotland's Highland Edge and the most noble shape in the landscape of my nature writer's territory. The day was cold and bright and white-washed high up, and more snow massing darkly in the north, and somewhere high on a long slope of big rocks and half-frozen snow, a face was watching me. My route skirted below the slope at first, but then I would veer away uphill towards the rocks, towards that steep and shattered boulderfield. It was that testing species of snow that sometimes bears your weight and sometimes bearhugs your leg up to the knee and it is as wearying on the mind as on the limbs. I felt the land with my stick, the one that was made on Mull out of cherry wood and topped with a curve of deer antler (its maker and I have no truck with fibreglass walking poles). I made slow progress and I walked looking down, which is not how I like to look and not how I like to walk. But whenever I stopped and looked up it was to see the rocks, and these beckoned. When I reached them I would sit in the sun and look at the land properly. And high on the bank of half-frozen snow, that face watched my every laborious step.

I had become half-aware of the face before I saw it, using that other sense the nature writer in me sometimes brings to bear on the moment, because hundreds of hours in this landscape and others like it have taught me to trust in it. It takes the form of something like a command from the very land itself to be alert, nothing more specific than that, except that because I know this hill passably well and know what I might realistically expect to see here, I have a shortlist in my mind at once of what it is that I think might be watching me.

Well, I would be a moving piece of its landscape a little longer, keep my head down while I climbed. As I closed in on the rocks and not far from some young hand-planted Scots pines that are slowly transforming the look and feel of the glen in a benevolent way, the snow was shallower and patches of heather and blaeberry showed where the wind had charged down the mountainside in the night, wind whose spoor still lay in long, even grooves that flicked up at the edges against the rock. Snow is a vivid story-teller.

I threaded a path through the first rocks until I found a flat one shaped a bit like a table by some ancient glacial upheaval and worn smooth by mountain time and mountain weather; a favourite perch of mine, also of buzzards, ravens, kestrels, an occasional golden eagle, and the owner of that face that I think might be watching me. I sat and raised the binoculars to my eyes in a single accustomed movement. It is how I sit down in all the boulderfields of my life. And soon, there at the far end of the glasses, was a face looking back at me.

It was the colour of autumn and snow. Its eyes

were smouldery gold. Its ears were tall and wide for the size of the face, two bluntly pointed triangles set high and facing forward. The fur on its narrow forehead and between its ears stood on end, but lay sleekly aslant on its cheeks, and that was the autumn of the face. The fur was darker under the eyes but it lightened again across the long, tapering muzzle. But from the black tip of the nose, a band of bright white travelled back the full length of the face and down to the thin dark line of its closed mouth. The throat was white too, but looked duller in the shadow thrown by the head, and that was the snow of the face. Nothing else showed. All else was hunkered low across a small and shallow rock terrace thickened by snow. The face was a fox, a vixen's I guessed, alluringly beautiful, entrancingly wild.

There was nothing else but the face. Snow had blown into a low drift along a ledge and the rest of her was hidden by its outer edge. I wondered if she was sitting, standing, or lying on the in-facing slope of the drift. Her face was so isolated by the drift that there was no way of telling.

Then two ravens found her, and they told me. They had already taken possession of the north wall of the corrie where they nest in a buttress every year, and even though January still had a week left to run and the snow was preparing new siege tactics with which to assail the ancient glacial battlefront of the Highland Edge, they had nesting on their mind, for they are among the very earliest of the mountain's creatures to stake their claim, and their dander was well fired up already and there is nothing on the mountain quite so methodically, determinedly hysterical as a late January raven with its dander aflame, nothing that is, except two ravens dandering in tandem, and amounting to rather more than the sum of the parts; dander squared.

They were cruising the ridge, loudmouthed and brash as Jets strutting their stuff to a Bernstein score, punctuating the libretto of nature's morning with random black oaths, back-flipping in synch and giggling through their practised choreography, the most amiable thugs on the mountain.

The vixen tilted her head up and left and slightly over her shoulder to assess the sound, the fuss, the familiar intrusion. The ravens were two hundred feet up and a quarter of a mile away. She looked back to re-assess my stillness, my silence, and when she turned again to the ravens they were line-astern and diving and she was in their line of fire, the target. She flattened, her head momentarily disappearing, and I knew then that she was lying on the snow. The ravens' dive bottomed out inches above where her head had been. Then they curved up, flipped over and were prepared at once to dive again, but in those few seconds the fox was on her feet and as they dived she stood on her hind legs to meet them with open jaws.

This was an event that was not in the ravens' script. They split left and right away from the fox, climbed again each in its own upswept curve at the top of which they seemed to confer, then each dived back down the very line of their ascent to come at the fox from opposite directions, but by then there was no fox. Neither they nor I had seen it move. The ravens perched just above the snowdrift, and if I were to hazard a guess at their demeanour I would say it was a state of confusion, rather like my own.It did not surprise me that I couldn't see the fox from where I was but I was surprised that the ravens couldn't. They looked in every direction, including up, as if it had perhaps taken wing. But it had simply vanished, that old wolf trick by which when they don't want to be seen, you don't get to see them.

The ravens departed in silence.

The mountain held the knowledge of the fox's disappearance in silence.

I sat on and watched and listened in silence.

I thought about that disappearance, and then I wondered if the snow might tell me how it was done. So I began to climb up through the boulderfield, heading for the fox's snow-drifted ledge and kicking good steps into the snow that I might have to use on the way back down, this with one eye on the lowering, advancing snow cloud that might complicate life on the mountain over the next hour. As I climbed I nursed a theory, one based on an old encounter with a different fox on this very slope of this very mountain, and snow was involved then too.

This is my kind of mountaineering. My interest in climbing a mountain to pronounce it climbed has long since waned, and now I climb so that I might unravel one more of the mountain's secrets. I climb with a particular purpose in a particular part of the mountain. You could say that I climb the mountain the way a fox does.

When I reached the ledge I found I could walk straight on to it from the open mountainside. The snow had piled against its outer edge, and thanks to an overhang it was much shallower along the inner edge against a low cliff. I was looking for a story to read in the snow but there were no fox tracks here, so however the fox had left the ridge it had not come this way. I edged further along the ledge keeping close to the rock, distrusting the outward edge, crouching eventually as the overhang imposed. There was no mistaking the place where the fox had been lying up, for there was a marked depression the length of its whole body. Beyond that, its footprints were everywhere, a crazy paving of footprints, but none of them led towards the far end of the ledge, so it had not gone that way either, and it had certainly not gone straight down. That only left the rock face, and that chimed nicely with the theory I had been nursing all the way up, the one prompted by that old encounter here with another fox.

Do you see the largest boulder down near the foot of the rockfall, the one the size of a small house? It happened there, and I had turned in to its leeward side to shelter from a sudden blizzard. I found a comfortable little ledge a few feet off the ground that ended in a right-angle of rock, and the angle of the two walls shut out the blizzard. It was done in about half an hour, but I was enjoying the seat, and sometimes stillness on a mountainside is more productive for a nature writer than movement. Sometimes I find I can slip into the mountain's mood of alert stillness, and again, there was that sense of a command from the land itself to be alert myself. And this can only be done alone. Climbing a mountain with other people has its own pleasures and rewards, but it is a different thing. Alone you can give the mountain your whole attention. There are no tribal loyalties to blunt the edge of what you might experience or limit what you might see. And in response to that command that is neither heard nor seen nor felt (yet somehow there is a prompting presence), I leaned forward to look down over the edge.

Less than six feet below me, on a little couch of heather, there was a fox. It lay on the heather like a cushion, its body tightly curved and its luxuriant tail draped across its muzzle, and in that attitude it had gone to sleep. My first reaction was to lean back, to withdraw my head from its line of sight if it should suddenly awake, but then, of course, I wouldn't be able to see it while it slept. And then (as I fancied it) the voice of the mountain:

The fox does not know you are here.

So I leaned forward again, inched my way into a slightly more comfortable posture, and just watched. Its flank moved to a slow rhythm. The snow was falling again but much lighter now and I watched it touch and melt against the vivid autumn-shaded fur. But where the curve of the tail lay against the fox's body, a thin lace of snow was lying. Over several minutes I watched that lace grow and thicken until it was a wedge of snow about two inches deep and as wide across the top.

A wild mountain fox's midwinter daylight doze was never going to last long. It was suddenly awake and instantly alert and on its feet, and of course it saw me looking down from just beyond touching distance. I was so sure its next move would be to jump down onto the snow and head off up the glen at a run. Instead, it turned in towards the rock and vanished somewhere beneath the very ledge where I was sitting. A few seconds later it reappeared fifty yards up the hillside where it stopped and looked back at me from the partial cover of smaller rocks. Then it ambled away up the glen and I watched until I lost it in the folds of the hillside. I clambered down to the heather couch to find an explanation for its vanishing trick. There was a crack in the wall of that boulder into which the fox had jumped from a standing start, flattening itself as it landed. By scrambling up on top I discovered that the crack went all the way through and that the fox would have emerged on the high side of the boulder facing the mountain and jumped a few feet down on to the snow from there, knowing it would be hidden from my sight until it was well away from me. All this was written in the snow.

◉ ◉ ◉

SO THAT OLD OBJECT LESSON in the degree of intimacy with which a mountain fox knows its mountain was now reinforced on the ledge of the fox-face and the ravens, for it seemed that here too the fox had sought to outwit both the ravens and me through the very rock of the mountain. And here too was the story written in the snow, the track of a four-footed leap back under the overhang where the snow was blown thinly, and something like a cave less than two feet high had offered black sanctuary. The small torch that lives in my pack showed only that the hole bent away from the ledge, and beyond that there was nothing to indicate how deeply or how high it penetrated into the rock face. Perhaps the fox was still in there, smelling me, listening to me, safe in the knowledge that none of the creatures that look and smell like I do can follow. And I wondered if the nature of this particular flank of this particular mountain had nurtured a singular family of foxes that generation after generation passed down the secret ways that thread the insides of the big rocks and rock faces, arteries of lifeblood for a fox that doesn't always want to be seen.

I climbed higher.

I sat again.

I waited.

Then I waited again.

I scoured the whole mountainside with binoculars, lingering over the ridge above and the corrie headwall where the ravens nest. Nothing moved except the snow cloud which had advanced and lowered and now darkened the day and whitened the whole glen, and that finally persuaded me to abandon my vigil, so I retraced my steps.

The new snow dragged down an early dusk with it, obliterated every trace of the mountain other than the next twenty yards and even my careful uphill steps had begun to blur around the edges, then to fill in and finally to vanish as if they had never been, but down by the burn the path that led towards the lower forested slopes was still well enough defined.

There was also a fox track and it was using the path where the going was easiest, and I know a Norwegian forest where winter wolves do that too. In this kind of snow, the footprints would survive no more than a few minutes before they were buried. This fox was somewhere just ahead of me. I picked up the pace and headed for the trees. January is the time to be out at dusk on the low slopes of a mountain like this one if you like foxes. It is the mating season, and the season has its anthem, the nearest thing Scotland has to wolves howling, a wailing scream with the very breath of the mountain night in it.

I followed the fox track wondering if this was the same fox and if so, how it might have got up through the rock face and then down below me on the mountainside without me seeing it. It's one of those questions I didn't waste much time on because I have no idea and no way of finding out, and besides, I am firmly committed to the ideal that all wild animals should confront me with mysteries from time to time so that I might embellish my admiration for them. Wolf, whale, eagle, swan, skylark, otter, badger, fox; it is true of all of them. It was darker in the forest and the tracks had dived into a black turmoil of spruce branches too thick for me to follow.

What are you doing? Where are you going?

The snow thickened again. The mountain was no longer a place to linger. The fox knows how to live here through the bleak midwinter; I do not. I followed the white curve of the track, my footfalls a companionable rhythm to the snow's deluge of quiet. I passed a solitary cottage beyond the edge of the trees, a light in the window. They will hear the foxes tonight from there in the snow-quiet of that room.

QUIET TONIGHT

Quiet tonight, not silent
– sideways snow
makes hasty whispers
at the window, not still
– rummaging winds
balloon the flakes
and rattle the black oaks.
What makes it quiet
is the pallid darkness
of the snow-stuffed sky
and the shy lull
that follows fox bark
after fox bark
after fox bark.

THREE

LINDISFARNE UNDER MIDWINTER STARS, a thin, gnawing east wind off the sea. When Margiad Evans wrote "the wind is a tooth in the breast" this is what she meant. The sky is ragged with clouds and fitful moonlight. The land lies so low against the sea that it could be Holland. In Highland mountains, with so little sky at your disposal and the rocks pressing in on your mind from both sides, such a night stuns the land with the illusion of profound stillness. On Lindisfarne, the entire world is astir. The sky with its unravelling patchwork clouds and hide-and-seek moon and stars is as mobile as it is vast. The falling tide has uncountable voices from far-out breakers to inshore white horses to the last sand-slapping tumble of spent waves and the hissy slither of inch-deep foam among seaweed and the myriad questing legs of wading birds.

The arrival of more and more birds as more and more beach unveils, birds in dozens and hundreds and accumulating into thousands, adds to the notion that even the land is jigging to the infectious rhythm of nature's island night. They come in low and loud (oystercatchers, curlews, redshanks, brent geese, six whooper swans) and they come in high and silent (plovers, a morose greenshank). From the sea too, the creamy voices of eiders drift ashore where they lay strange harmonies on the muted brass of the swans. Nothing sleeps at such an hour in such a place, for there is night feeding for all and a flickering moon for a candle to light nature's table.

Enter stage right from the shadows of the dunes where they have been toying with and terrorising the rabbits through the dusky hours, two foxes. On Lindisfarne the foxes' diet has salt with everything. The fur of rabbits on such an island is routinely drenched with wet salt winds. And when the tide falls, the foxes take to the beach and explore the rockpools for crabs and washed up fish, but when the beach is as full of birds as this, and nightfall subdues their daylight-red coats to a dull and indeterminate grey, they are apt to set their sights a little higher up the food chain. The birds know they will come, of course, for they come every tide of every day and night, but the beached flocks display a fatalism based on safety in numbers, in much the same way that they respond aerially to the ripping flight of the peregrine. The theory is that the sheer weight of numbers and the chaos such numbers create in the face of a predator will defeat that predator more often than not, and if there is a casualty it is one casualty at a time and every other bird has survived, so the law of averages is stacked in their favour.

The foxes don't seem to feel the need for stealth at night, and trot along the edge of the water not a hundred yards from the nearest crowd of waders, while looking all around, sizing up possibilities. It may be (I have been told) that the birds'-eye-view of them in this light does not recognise them as foxes, but I am sceptical. I think that many thousands of years of evolution will have taken care of that. On the other hand, most beaches are not like Lindisfarne beaches, and most beaches don't have foxes, so it may be that there are exceptional Lindisfarne dimensions in the relationship between predator and prey species.

From where I am sitting at the top of the beach (and where I have been sitting since before the birds began to arrive and since before the foxes got fed up chasing rabbit shadows), and what with the moonlight and a good pair of light-gathering binoculars, the foxes still look like foxes to me even at two hundred yards. They look skinnier and longer-legged than the mainland mountain foxes I know best, but that could be just because they are wet most of the time, or perhaps (it's impossible to tell from here at night) they are last year's cubs out on the prowl and struggling through their first winter.

One fox has stopped, and flattens on the sand like a collie awaiting instructions. The other has trotted on as if it has not noticed. I suspect a ruse. Unease whispers through the flock like wind through a reed bed. My problem now is that the foxes are a hundred yards apart and I can't watch them both in the glasses. I opt for watching both without glasses. The walking fox stops suddenly, sits on its haunches, yawns, looks round, looks bored, scratches its nose with a hind leg.

The other fox is still motionless on its belly, but is it my imagination or has it squirmed ten yards closer? The sitting, scratching fox stands again and begins to trot towards the birds, looking for all the world as if it has still not noticed they are there. The flock's tolerance snaps, there is a mass movement to turn and fly away from the trotting fox which means that the birds furthest from it are the last to fly and as they turn and shake out their first few wingbeats they find the second fox almost in their midst. Even as it leaps off the ground to snatch a bird from the air it is clear that the birds have got away with it this time, with about a yard to spare.

It is the first time I have seen foxes hunting as a pair, deploying tactics. The night is chaotic with wings for several minutes, the air alive with milling winds, but surprisingly few bird voices. The foxes have come together on the sand to dance around each other for a few moments then they trot back to the shadows.

I am a rare visitor to Lindisfarne. It is an island with a very persuasive magic, but my island-going inclinations tend to lure me north and west rather than south and east. So I don't have enough of a picture in my head of the way the island foxes do business. I do know, however, that like foxes everywhere, they suffer at the hands of some of the human natives. There are communities all over the country that make a lot of money from visitors who come for the landscape and the wildlife – for nature. Accommodation adverts for Lindisfarne always mention the birds, and, the foxes. But I have yet to find a community the length and breadth of the land that is without a hard core that will make an exception of the fox in their relationship with nature. The following is from a fine little book called *Lindisfarne Landscapes* (Saint Andrew Press, 1996) by Sheila Mackay, a friend of many years who for a while was a part-time resident of Lindisfarne:

The afternoon of New Year's Eve had been arrestingly dramatic. Walking on the Snook I noticed a man perched high on Primrose Bank, dark-sweatered, clutching a shotgun, immobile, with his head turned towards Snook Point. Watching, concealed, I wondered what would happen and could scarcely believe my eyes when a posse of maybe twenty men and older boys appeared, strung out in a line from Primrose Bank to the North Shore, each with a shotgun at the ready.

Passing Snook House garden, they saw me and their greeting dispelled my apprehension: they had seen four foxes but killed none. The foxes had to be flushed out and killed, they said, for they were plundering the refuse bins and attacking the village hens.

⦿ ⦿ ⦿

EVEN HERE, I THOUGHT, even in this place so given over to nature, so shaped and reshaped by nature, so defined and redefined by nature, so dependent on nature for its bounties and (as far as the people are concerned) for a living… even here, there is intolerance for foxes. No matter how many mice and rats the foxes will kill, no matter how much joy the presence of the foxes brings to so many people, it's the same old damnable story.

I am very fond of Lindisfarne. It has a unique atmosphere. It is an astounding theatre of nature even at night. Especially at night. Its relationship with foxes is no worse than anywhere else, but in the eyes of a nature writer it is perhaps just a little more disappointing. Every time I hear about retribution being visited on foxes because they attack hens, I remember the words of the old Scottish naturalist David Stephen: "I never yet saw the fox that had the key to the hen house in its jaws."

It has taken perhaps ten minutes for the sands to start filling up with birds again, although the swans did not fly, but only walked away a few yards, muttering querulous pairs of muted-brass semi-quavers in their throats. I have yet to meet the fox, no matter how hungry, that that had any appetite for a head-on confrontation with a healthy adult swan.

FOUR

St Margaret's Loch lies at the foot of Arthur's Seat, a miraculous little mountain within sight and sound of Edinburgh's Old Town. Its situation at the heart of Holyrood Park may be tame, and although its swans (and its geese and ducks and gulls) will almost all take food from your hand, they are still wild in their seasonal flights and in their characteristic behaviour. On a midwinter night, the mountain is a black mask and the city seems suddenly far off. The loch is hard ice, and draped in a blue-black sheen of bounced light from the city's night sky. About sixty swans are gathered here, and clustered around them are hordes of the lesser fowl of the night, grateful for the restlessness of the swans that constantly break down each new creeping frontier of ice to keep a corner of the loch ice-free.

A night like this at this time of year can mean anything up to eighteen hours of darkness to stand through, to stare through, to get through in a one-footed doze, to bicker through and reassert a claim on a small space of the water or the ice. Yet these birds are the lucky ones, for the city feeds them by day. Carfuls of Samaritans bring bread and other scraps. Most of the birds will eat well enough to thole the nights, a fact not lost on the city fox which traverses the mountain in the evening from a daytime refuge under a shed in the fag-end of a Duddingston garden. Nor is the fact of the ice lost on the fox. Where ducks and geese can stand and walk, so can a fox.

He is a hunter, and cold and hunger only serve to keen his hunter's instincts and embolden him. He will know there may be a gauntlet of swans to run, but he knows too that if he watches from a distance, prepared to wait out as much of the night as it takes, sooner or later the restless throng will carelessly expose one unwary duck.

FOX

A distant bell sounded 10pm, chimes that wavered through the vagaries of the wind, now bright, now muffled. My way was up through the whins with a torch for the pitfalls. Eight years as a newspaper journalist here had given me a back-of-the-hand familiarity with the middle of Edinburgh and these were my most favoured of all the city's acres.

Above the nearest whins the track dipped into a hollow that shut out the city and darkened the night with deep shadows. It was there I met the dog fox, there that he turned from me, disappeared at once heading for higher ground, but I heard his feet thumping frozen grass, then higher up, frozen snow. There I saw him again, trotting and relaxed again, and with his moonshadow for company. He stopped once to look back, then changed course onto a long downhill curve that headed for the loch. I took a detour of my own so that I came on a small outcrop that looks down on the loch, and there I sat and prepared to shiver for a while.

After about twenty minutes, I saw him walk slowly along the lochside path at the foot of the mountain, moving away from the swans, intent on a scattered group of ducks standing one-footed on the ice towards the far end of the loch. He took a few tentative steps onto the ice, then a dozen more purposeful paces. Then all hell broke loose.

Two swans rose from water onto ice, and in a running, sliding fury with necks stretched straight and low and wings threshing the ice, they drove at the fox from fifty yards, scattering ducks, geese and gulls in squalls of protest. What the fox would have seen as he turned his head at the first explosive cracking of the ice was this:

Two wingspans of around seven feet, and wingtip to wingtip, charged at him like a breaking wave, a wild white wave tumbling across the ice and rendered all the more eerie by the back-lighting of the city and their weirdly mimicking ice-defused reflections. The swans never really became airborne; the wings were deployed to make sound and fury and awful spectacle, and the bowsprits of their heads and necks were unmistakably aimed at the fox. The fact that the birds were running on ice added chaos to the charge and that only added to its ominous edge. And all this was happening at the fox's eye-level. It is hard to imagine he ever saw a more terrifying sight in all nature.

Considering the underfoot conditions, his retreat was surprisingly deft and sure-footed until he was within a yard of the shore when he lost his nerve completely, crashed chin-first onto the ice, recovered and clambered awkwardly onto the path while the swans subsided onto the ice but careered on another twenty or thirty yards completely out of control, all decorum shot to hell in a virtuoso routine of panto slapstick. By the time they were on their feet again, the fox was a shadow among shadows.

We retired, the fox and I, to such food and shelter as we could glean from the night city, and (from our very different points of view) to dwell on the vigilance of the white sentries of the winter night under an urban mountain.

FIVE

Flanders Moss is one of those national nature reserves with a car park, a well-made circular path, an observation tower, a couple of judiciously sited benches, and a series of interpretation boards that depict and spell out some of the peculiarities of the mysterious denizens of Europe's largest raised bog. It's a claim to fame of a kind.

So you will be introduced to the world of toads, frogs, newts, water-bugs (not an actual species but a generic term I have just made up to hide my ignorance of water-bugs), dragonflies, damselflies, dragonfly-and-damselfly-slaying birds (tree pipits, reed buntings and redstarts, for example), drumming snipe, piping curlews, mosses and water-loving plants, winter geese and whooper swans and short-eared owls and woodcock, a day-flying moth (the argent and sable) and a moth with no wings (the female Rannoch brindled beauty), vagrant otters using man-made ditches to travel between watersheets, red deer and roe deer. But they make no mention at all of the common gulls and the foxes.

I am not an enthusiast for countryside "interpretation". The very word seems to me to imply, "you don't know what we know so we're telling you a little bit of what we know in simple words we think you will understand, and we'll tell you some of what you should be looking for but not anything so important we have to keep it a secret from you". But at Flanders Moss, at least the touch is light, the language is unpatronising and avoids the much-loved tendency of nature interpretation board writers to use tabloid-esque jokiness lest you might be under the mistaken impression that nature is important. The path is well-made and evolves into a well-made boardwalk that lets you look at the raised bog and its unique wetland community in the eye without stepping up to your waist in glaur or simply disappearing into it forever. But the signs still don't mention the common gulls or the foxes.

FOX

So this is what the signs don't tell you. The common gulls turned up one afternoon in early February, a small advance party of eight, speculative nesters I guessed, having drifted far inland via the Firth of Forth, Stirling, and the broad plain of the Carse of Stirling, above which lies the Moss. I say above, I mean about twenty feet above. It is noticeable as soon as you leave the car park (which is at the same level as the fields of the Carse) that a short path rises up at once through a strip of handsome birch wood before it bursts into the wide-open, level sprawl of the Moss, which you now see lies like a glorified mezzanine of nature slung between the Carse and the foothills of the Highlands. Beyond these, an arc of mountains from Ben Lomond in the west to Stuc a' Chroin in the north scrawls across the sky the unambiguous signature of the Highlands. The next piece of flat country is Rannoch Moor, fifty miles further north, 1,500 feet higher than this, and as resolutely Highland as the Moss is Lowland.

So it was into this wide waterworld that the gulls wandered and it was there that I met them that February afternoon, watched them drift low over the Moss, circle twice on gliding wings, then land on one of the Moss's countless pools, the nearest one, as it happened, to one of the benches where from time to time I like to sit and write and look around at the wild world between sentences.

At first I was wary of disturbing the birds, but after a few days of afternoon visits to the bench they treated me with indifference and I could watch them and photograph them and write them down at my leisure. And then there were six.

My first thought was that they had just moved on. It happens. Then one late afternoon, just as the light was going and I was taking my leave, I saw a dog in the distance on the footpath. It's not unusual. People like the easy nature of the path, the beauty of the surroundings, and they bring their dogs with them. Two things made me wonder about this particular dog: one was that it was an odd time of day to walk a dog out here, just as it was getting dark, and the other was that it did not appear to have a human accomplice. So I had a quick look with the binoculars and what snapped into focus was a dog fox.

FOX

I was a long way from the gulls' pool and heading further away, and towards the birch wood and the car park. The fox was coming the other way, and loping along the boardwalk as if it had been provided for foxes rather than people and their dogs. I found a relatively dry patch of ground beside the boardwalk, stepped off carefully, and tested every step of the few yards to a sapling birch a little less tall than I am, and stood behind it. In summer its leaves would have screened almost all of me, in February its skinny branches might just help to break up my shape in the eyes of the fox if he wasn't in a particularly noticing mood. It was a forlorn hope, for the fox is the most noticing mammal in all of wild (and wolfless) Scotland. Besides, he would find my scent waiting for him as soon as he drew level with the birch, and if anything his awareness of scent outsmarts his awareness of sight.

He duly arrived without breaking stride. I could hear the rhythmic patter of his feet on the boardwalk. About twenty yards away he looked straight at me. His stride never faltered and he walked straight past but his eyes never left me until he had gone a dozen yards further along the boardwalk. Then he stopped, turned and stared at me for several seconds, and then he turned and walked on, resuming the same resolute pace as before. I watched him fade into the gloaming, as I have watched so many foxes fade into so many different landscapes. And two days later there were four gulls, and I began then to put two and two together and make four fox meals of the ones that had disappeared.

Nothing changed until early April, and although I made several afternoon visits that lingered until almost darkness I didn't see the fox again. The gulls were settled and apparently untroubled on the same pool. And then there were two.

And then I walked out one bright spring afternoon that was giddy with skylarks and tree pipits. Halfway to the bench by the pool, one of the gulls rose and came straight towards me, circled low overhead calling loudly in tones that said "piss off" in any of the languages of nature. I walked on until I could see the pool, and there in the binoculars was the head of a sitting bird deep in a bush of thick reedy grasses. I wished them good luck, and headed for the far side of the reserve, and there on the path I found two fox scats.

I settled down to wait for the dusk. The denouement came while there was still just enough light in the west to see clearly anything that crossed that patch of sky. There was a sudden clamour of frenzied bird cries. I looked up to see both gulls low over their territory, and I headed for the pool. I was still a hundred yards away when the commotion reached fever pitch and both gulls were constantly diving down towards their pool, screeching as they dived.

The fox was wet when I saw him, wet from his short swim to the nest, and my best guess is at he had just cleaned out the nest of its eggs then swum ashore again. He cut away across the bog, not putting a single foot wrong amid so much underfoot treachery, an angled run that met the path a few yards further on. There he stopped once more and looked back at me in a perfect reprise of the attitude with which he had studied me behind the small birch tree. Then he looked back at the pool with its hapless gulls still circling, still crying. Then he turned and trotted off back towards the birch wood.

The next time I visited the Moss, there were no gulls, and whether they too succumbed to the fox as they tried to nest again, or whether they had simply given up and moved on, I have no way of knowing. They must have thought that their deep pool with its tiny, densely-grown grass islands was a perfect nest site, but they were not the first and they will be far from the last creatures to underestimate the ingenuity of foxes. It is easy to take sides in circumstances like these, and there are those who will consider it inhuman not to. But it is nature's way that prevails here, and nature has endowed the fox with many uncanny survival skills, only one of which is that of an enthusiastic swimmer. The fate of the common gulls of Flanders Moss is just the way it is sometimes. You and I, being members of the one species that has shown more vicious animosity towards the fox than all the predators of nature combined… you and I have no right to judge.

SIX

THE TROSSACHS IN MAY SUNSHINE. You can still see the places high up where the cliff face burst open, where slabs of rock tumbled down the mountainside and smashed into boulders as they fell. In this kind of strong sunlight you see the broken cliff face's tinfoil shimmer, for the geological wound is still too new to have acquired the grey patina of age.

An old and sparse mountain woodland of mostly oak and birch, browsed by more than 150 years of sheep and deer beyond any hope of significant natural regeneration, had stood in the path of the rockfall. Some of the rocks simply broke through the wood. Trees were uprooted or broken at the waist, or had great limbs torn away. You can still see the casualties. But other trees stood firm, slowed or halted the progress of the fall, and when the dust and the appalling noise settled, a new and precariously poised boulderfield began to adjust to life several hundred feet lower down the mountain from the cliff that had started it all.

So it was a boulderfield with broken, wounded and dying trees, or at least it was at first. But nature is not one to miss an opportunity even if it is one born out of such destructive upheaval. For the boulderfield's sudden and brutal arrival has been the making of the wood, or rather the re-making of the wood. It is such a steep place and so treacherous underfoot that grazing animals rarely set foot there now. Thus, nature swarmed all over the rocks and mangled trees and smothered them with new possibilities.

Other trees rushed into the new vacuum, some the progeny of the crippled survivors of the rockfall, some borne on the wind, some carelessly transported by birds. Woodland plants followed, wood sorrel, wood anemone, wild hyacinths. They all found niches among the boulders and the dead and dying and recovering and wound-licking trees, and the old wood found new strength and spread higher and wider across the face of the mountain. So now you find not just oak and birch but also young rowan, holly, ash, aspen, willow, juniper, pine.

I got to know the place first when I was watching its badgers. An old hand in the glen told me about them, and about the rockfall. The badgers were attracted westward from a sett a few miles along the glen by the huge chambers that were created where boulders had come to rest in that ungainly chaos. The badgers scarcely had to dig out their setts at all. They adapted to their new surroundings by becoming skilful mountaineers, front-pointing up almost sheer rock faces with some style. It has always seemed to me to be a part-time sett, perhaps an outlier to be used when the sett to the east is overcrowded. But the fox den is full-time.

It was while I was watching the badgers that I found the foxes. I was walking in to the sett one spring evening in the last of the light. I had stepped across a skinny burn just above a little waterfall that must have masked the sound of my feet, and rounded one more rock when I simply came face to face with a fox.

We were both on the same narrow path, with no room to pass each other. We both stopped and stared at a distance of about twenty feet. We both assessed the situation. We both thought in our own way, "What now?"

He was a dog fox, returning from a successful hunt. A small rabbit hung in his jaws, limp as an empty sock. His coat gleamed foxy red, darkening towards his hind quarters. His ears and lower legs were black, the tip of his tail and his chest white. I thought it odd that he had not turned and run.

Where are you going? What are you doing?

I made it easy for us both. I started to retreat, walking carefully backwards and speaking to him in what I hoped was a calm and benevolent tone of voice, while he watched with his head slightly on one side. After a few yards, I turned and walked away from him towards the edge of the wood. I then did something very specific simply because a week earlier I had seen a badger do it. I rounded one of the huge boulders that had a thick mattress of heather and moss on top, then climbed its "blind" side until my head was level with the top.

My progress from there to a flat-out prone position on the springy green mattress was infinitely more ungainly than the badger I had watched, but I made it without too much fuss and I was now ten feet higher than the path, hoping that would be enough. Here, the badger simply lay down and watched the wood. I now did the same thing.

There was, of course, no sign of the fox. But there was an hour of usable light, and if I lay silently and still, who knows?

A mistle thrush was singing nearby. I have come to associate two birds in particular with the environs of a working badger sett. One is the robin, which seems to delight in feeding on badger spoil heaps and into the very mouths of entrance holes. The other is the mistle thrush. I once timed one as it sang – just over two unbroken hours of song, song that lay on the still woodland air like a cloud of silver smoke. The musical invention of the bird is uncanny and kaleidoscopic in its colours and patterns, and its music also offers the perfect antidote to the occasional tedium of a badger watch when nothing at all happens.

I lay on the rock listening, and the fox suddenly walked back along the path going the other way, and minus the rabbit. So somewhere nearby there was a vixen with cubs. So I changed my working routine at the sett. In an effort to encounter the fox more often out in his territory, I started to work the hill above the wood in the afternoons, and come down to the badger sett in the evening. Three weeks later I saw him walking along the top of a drystane dyke, pausing often to scent down in among its stones. Suddenly a pair of wheatears were flying around his head. I put the glasses on him, just in time to watch him devour a clutch of nestlings one at a time from their nest inside the dyke.

He jumped down from the dyke and headed uphill towards a deer fence, turned and followed it east until he came to a stile, nothing more than a few parallel straps of wood like a wide ladder, but completely vertical. He ran at the style from about five yards away and leapt, hitting it about halfway up and scrambling over and jumping down the far side in a single fluid movement.

"Hmm, you've done that before," I said aloud to no-one at all.

I did not see his cubs that year, but the year after, I heard that four had turned up dead on a nearby eagle eyrie. I mentioned it to the old hand in the glen who simply said:

"Ah the balance of nature, eh?"

So I quoted Snoopy to him; you know, Snoopy from Peanuts:

"Those who believe in 'the balance of nature' are those who don't get eaten."

AFTERWORD

THE RED FOX (*Vulpes vulpes*) is widespread throughout the northern hemisphere, and ranges from the edge of the Arctic to African deserts. In Britain its population is thought to be around a quarter of a million, and many foxes have coats of deep reddish-brown with white underparts and a white tip to its brush, but the colour and markings can vary widely. Black foxes, although very rare, have, as you can well imagine, a folklore all their own.

The fox is an endlessly adaptable animal. Its diet, for example, is almost as easy to define by what it won't eat as by what it will. In the wild, rabbits are a particular favourite, and it will take almost any small mammals, including mice, rats, voles, moles, ground-nesting birds and their eggs (which brings it into conflict with keepers), farmyard poultry and weak or dead lambs (which brings it into conflict with farmers), berries and fallen fruit, even the odd piece of dead fish that an otter might have discarded. Its spread into towns and cities in recent decades has introduced its opportunistic temperament to a wide range of discarded human food up to and including curries, kebabs, cheesy chips, burger buns and white bread, none of which have done the animal any favours.

Its habitat ranges from urban streets to sea shores and from farmland to woods, commercial and natural forests, moors, hills, and the highest mountains (it is regularly seen hunting birds like ptarmigan and dotterel on the Cairngorms plateau at four thousand feet.)

Some biologists have suggested that the urban fox is evolving into a different sub-species given the very different nature of its habitat and the demands put upon it, the effects of that unsavoury diet, and its much shorter lifespan. The average lifespan of a fox in the wild is about four years, but in towns and cities it is less than two. The evolution of the urban fox does not yet include road sense and urban traffic kills huge numbers every year. In many parts of the country the urban fox is a leaner, rangier, longer-legged animal than the hill fox.

In the wild, the fox den is in a cairn of rocks, or more often in an underground hole of some kind – an enlarged rabbit burrow or a deserted badger sett, and even an occupied badger sett where they seem to reach an amicable accommodation with their hosts. In the city, almost any kind of secret nook will serve; many live beneath garden sheds.

Adult foxes mate in January or February, which is when they are at their most vocal. The dog fox has a triple bark, the vixen's is hoarser, but the defining voice of the mating season is that hair-raising wail-cum-scream. Most cubs are born in March or April, the later the further north. Typical litters are between two and five cubs although up to ten is possible. Cubs are born blind and with grey or brown fur and when their eyes open after about a week they are blue. They stay in the den with the vixen for a month while the dog fox hunts for all of them, and are introduced to the over-world gradually. Through the summer, cubs and adults make more adventurous trips together and the cubs usually move out for good in the autumn.

The hostility of farmers and grouse moor keepers has devoted inordinate resources to the perfectly futile practice of killing foxes. They are routinely shot, (especially at night by "lamping"), snared, gassed, poisoned, and otherwise hounded to death by legal and illegal means. Hunting with dogs was outlawed in 2005, which does not mean that hunts have stopped going out with dogs chasing a lure in known fox territories.

The fox's impact on rural life has been over-stated for centuries, and used as an excuse to prop up Britain's favourite "country sport", which is killing foxes by fair means and foul. The fox's capacity to thrive despite their best efforts has won grudging admiration from many a farmer and keeper, but in very few cases does that admiration extend to just leaving them alone. Yet among those very few cases, there is sound evidence that where foxes are left alone they are much less troublesome to farms and estates. For there is also evidence that foxes respond to "traditional" estate harrassment by increased breeding, which in turn creates a surplus of young foxes without territories ready to move in as soon as culling empties an established fox territory of its resident family.

The twenty-first century should be no place for what is essentially Victorian land-use philosophy. We should do better than to consign and condemn, under the loathsome catch-all of "vermin", everything in nature that is inconvenient for a particular human use. It is not so long ago – uncomfortably within living memory – that animals like the otter, the badger and the pine marten were routinely killed as vermin. Now they are protected by law. Surely our twenty-first-century awareness of how nature's scheme of things works should be willing to accommodate the red fox, and to offer it the same protection.

Its human enemies have been getting away with murder for far too long.

ORGANISATIONS DEDICATED TO red fox conservation include The Fox Project and the National Fox Welfare Society.

www.foxproject.org.uk

www.nfws.org.uk

⊙ ⊙ ⊙

JIM CRUMLEY IS A NATURE WRITER, journalist, poet, and passionate advocate for our wildlife and wild places. He is the author of more than twenty-five books, and is a newspaper and magazine columnist and an occasional broadcaster on both BBC radio and television.

He has written a companion to this volume on the barn owl, with further ENCOUNTERS IN THE WILD titles planned. He has also written in depth on topics as diverse as eagles, wolves, whales, swans, native woods, mountains and species reintroductions.

⦿ ⦿ ⦿

Published by Saraband
Suite 202, 98 Woodlands Road
Glasgow, G3 6HB
www.saraband.net

ISBN: 9781908643759

Printed in the EU on sustainably sourced paper.
Cover illustration: Carry Akroyd